This book came about because one day a boy called Sam Green saw some Soay sheep and was curious...

This is a Soay ewe called Daphne.

There are less than 3,000 pure breed Soay sheep in the world. Soay are probably the oldest kind of sheep in existence, forerunners of all the millions of sheep we have on farms today.

But where did they come from? No one really knows. This, and the fact that they are so old, makes them particularly special. Scientists study Soay not only to try to find out where they came from but for all that they can tell us about other animals.

Until just over one hundred years ago the only place in the world where they existed was on one tiny rocky island called Soay, one of a small group of islands called St Kilda.

Soay sheep on St Kilda.

St Kilda is fifty miles out into the North Atlantic ocean from the Outer Hebridean islands of Lewis and Harris, off the north west coast of Scotland.

The islands of St Kilda are the remains of a long extinct volcano, and one of the wildest and most remote places in the British Isles. In winter the winds on St Kilda can reach more than 200 miles per hour. Because of this and the frequent storms it is only possible to get there by boat for a few weeks in the summer.

The islands rise hundreds of feet straight out of the ocean on huge grey cliffs. On top of the cliffs the ground is rocky and hard. Not a tree or a bush is to be seen. The last people to live on St Kilda left in 1930 and today the only inhabitants are tens of thousands of sea birds – mainly gannets, puffins and fulmars – a rare type of field mouse and the Soay sheep.

The sea around St Kilda can be treacherous, but local sailor Murdo Macdonald knows its waters as well as anyone. It takes eight hours to reach St Kilda from the west coast of the Hebridean island of Lewis in Murdo's small, twelve berth boat, the Cuma. If the weather is bad you won't get there at all, but if you are lucky and the weather is kind, as you near the islands you will be surrounded by whales and dolphins, the tips of their tails and back fins dipping in and out of the waves as they swim round and round the boat.

In this picture a seabird – a skua, also known as a bonxie – is attacking the person taking the photograph!

The Cuma approaches Soay Island, St Kilda.

For over a thousand years one hundred and more people lived on the biggest of the islands, Hirta. To survive, everyone had to work, the men, the women and the children. They lived off what little food they could grow, together with the island's wild plants and herbs, the seabirds' eggs and meat from the birds and from their sheep.

This photograph, taken in 1886, shows the women and children of St Kilda. Can you see they have no shoes?

Hunting seabirds was especially important to the St Kildans. Not only did they eat the birds' meat, they could sell the feathers and oil to people on the mainland. The feathers were used to stuff into pillows and the oil – this was in the days before the invention of electric light – was burnt in oil lamps.

The men of St Kilda with their arms full of seabirds which they caught on the cliffs (1886).

However, the winters, with their frequent storms and fierce, driving rain, were hard. Stuck far out in the Atlantic the St Kildans were often cut off from the mainland for weeks or months at a time. No matter how hungry they might become if their food ran low, or how ill they might be, they could not get a doctor or any other help from the mainland until spring. It was not surprising therefore, that about one hundred and fifty years ago, some of the younger people began to leave to seek an easier life elsewhere. As the years passed steadily more and more people left, until by 1930 there were only 36 left. That was no longer enough to hunt all the sea birds they needed for oil and feathers to sell to the mainland, or to spin and weave cloth from their sheep, to plant the crops and do all the other jobs that had to be done if they were to survive.

The men met each morning to decide which jobs needed doing that day. Some of them have no shoes either (1886).

So the remaining St Kildans reluctantly decided that they must abandon their island for ever. On the 29th August 1930 the last St Kildans sailed away never to return. The islands were left to the birds, the field mice and the Soay sheep.

Why are Soay sheep special?

A lot of things about Soay sheep remain unknown. What we do know is that about a thousand years ago, when Vikings and Norsemen first sailed their long ships around Britain and came upon St Kilda, they found the sheep already there on their little rocky island. Seeing the sheep on the island, they called it 'Sheep Island' – in Norse, 'Soay'. The name stuck, and from that day to this the little island and the little sheep which lived on it are both named Soay.

The island in the distance is Soay, named by the Vikings because of the sheep.

But if the sheep were already there a thousand years ago when the Vikings came, how did they get there in the first place? No one really knows. Scientists believe that the sheep were probably originally taken there by settlers some 2,000 years before the Vikings arrived, in what is called the Bronze Age. However, they could have been there for 5,000 years, or more. Some people think that they may even have been left over from the time before the sea separated Britain from the rest of Europe. Soay sheep remained almost completely confined to their one tiny island until 1932. Then, two years after the last people left St Kilda, some were moved to the biggest island, Hirta. Today there are about 1,500 Soay sheep on the St Kilda islands. No more can survive there because, with the harsh conditions, there would not be enough grass to eat. In the winter they will even climb down to the one little bit of shore and eat seaweed.

In 1963 a small flock of Soays was taken off St Kilda and brought to England to be studied by a famous scientist called Professor Peter Jewell at London University. As well as finding out about them and their history, he wanted to make sure that there was a proper flock of Soay somewhere apart from St Kilda in case something such as serious illness struck the sheep there and they died out. Today there are almost as many Soay sheep in England, Scotland and Wales as on St Kilda. There are also a few in America and Europe. All of them are descendants of the Soay sheep that were originally found on Soay island.

Soay are smaller than most other kinds of sheep. Most have horns, although there are some that never grow any. The ewes (females) have horns that are 18 to 20 centimetres (7 or 8 inches) long – that is about the same as the height of this book – and slightly curved. The rams (males) have bigger, heavier horns which curl round the sides of their heads when they are fully grown and a thick mane of wool round their necks.

A ram with big curled horns.

A ewe with a lamb.

Soay range in colour from very dark brown to very light fawn, almost white. Most have white patches on their stomachs, legs and backsides and usually have patches of white wool around their eyes and mouths as well.

Some of these Soay sheep on St Kilda are almost black while others are much paler.

Look at the mane round this ram's neck.

In this photograph you can see the white patches on the ewe.

Soay sheep shed their wool in the summer like deer shed their winter coats. Because of this you do not have to shear them but can just lift the wool off when they are ready to moult. Soay wool is very soft but also, as you would expect from sheep from such a cold and stormy place as St Kilda, it can be woven or knitted into very warm clothes which are perfect for winter.

When the wool first comes off a Soay it feels oily and smells sheepy. If you rub it between your fingers it makes them feel all soft and smooth. This is because of the amount of sheep oil, known as lanolin, there is in it. The lanolin helps to keep out the rain and keep the Soay warm and dry in the St Kilda winter. Before it can be spun the wool has first to be washed and cleaned to remove all this oil and other dirt. Then it has to be combed – or carded – using a special iron comb. This makes the wool soft and fluffy. It is only after all this has been done that the wool is ready to be spun. Once spun, the Soay wool can be woven into cloth or knitted into warm winter clothes.

Raw, carded and spun Soay wool.

Soay ewes normally have one or two lambs each, which are born in the spring. When the lambs are born they are very small but they quickly get up on to their feet and start to run about. The mothers are protective towards their lambs and look after them very well.

This ewe's lambs are just two days old. Ewes are very protective towards their lambs.

This ewe's lamb is called Gus. He was born only a few hours ago.

Soay lambs quickly learn to stand up and can run very fast by the time they are just a few days old.

Soay ewes live until they are fifteen years old or more. They can start having lambs when they are one or two years old and are able to go on having new lambs every year for most of their lives. A lot of Soay keepers only let their ewes have lambs once every two years so as to give them time to get strong and well again between having each set of lambs. Rams usually only live for about half as long as ewes.

This lamb is about a week old. Lambs do not have horns when they are born, but you can already see a dark patch on the top of his head where his horns will grow.

Soay sheep have to be very healthy to survive on St Kilda. As a result they do not get many of the diseases that ordinary farm sheep often get. Really they are still wild sheep and in many ways are more like small deer than ordinary sheep. They are very shy. They have thin, but strong legs and small feet. They can jump and run just like deer and can be very hard to catch. Soay sheep will sometimes jump right over the top of a farmer's head when he is trying to catch them and they can easily jump over most fences. If they are taken somewhere they do not like, or are separated from the rest of their flock, they will jump over the fence and run off. That is what happened recently when a farmer bought some Soay sheep, took them back to his farm and put them into one of his fields. After exploring the field for a few minutes they decided that they didn't like it. So one after the other they all jumped over the fence and ran off. That farmer never saw his Soay sheep again!

Nettle

Once Soays get to know people they can become very tame.

One day when some boys from a boarding school were out for a walk they came upon a little Soay lamb all by himself in a patch of stinging nettles. Neither the lamb's mother, nor any other sheep, was anywhere to be seen. So the boys picked the lamb up out of the nettles and took him back to their school. By the time they got there the little lamb was bleating pitifully. So the boys found a box and some nice fresh straw, put the lamb into it and hid him in a cupboard where no on would see him. Next they hunted about, found a baby's bottle and filled it with milk. Then, still making sure that no one saw them, they went to the cupboard, took the lamb out of his box and, holding him in their arms, gently pushed the end of the baby's bottle into the lamb's mouth. Immediately the lamb stopped bleating and started to suck eagerly. For the next week or two they continued in much the same way, hiding the lamb in different places around the school and feeding him whenever they could. Because of where they had found him they called the lamb Nettle.

But as the days passed it became harder and harder for the boys to keep Nettle a secret. He had started to try to follow them around and even seemed to want join in their games. Soon the school holidays would begin and, as none of the boys had anywhere suitable to keep a lamb at home, there would be no one to look after him. So they reluctantly decided to tell their teacher about him. When the teacher saw Nettle, and how lively he was, she congratulated them on how well they must have looked after him. Nevertheless, Nettle couldn't live at the school any longer and a proper home would have to be found for him. He was getting bigger and it was time that he started to live in a field with other sheep, where he could run about and play as they did. Luckily the teacher knew a lady who had a small field beside her house and wanted to start a small Soay flock of her own. So she and another teacher put Nettle into their car and drove to the lady's house. When they arrived they put Nettle into the field beside the house, shut the gate and went to speak to the lady. A few minutes later they were all sitting in the kitchen talking when there was a little bleat at the open door and there was Nettle looking at them. He was so small that, left on his own and wanting to know what was going on, he had squeezed under the field gate. For a moment Nettle looked at each of them in turn, then he walked over and sat down beside them as if ready to join in their discussion. Sophie, for that was lady's name, quickly agreed that she would be delighted to have Nettle as the first sheep of her new Soay flock.

Sophie feeding Nettle from a bottle.

By now Nettle was so used to living with people that even months later, after two more Soays had come to live with him in the field beside Sophie's house, it still often seemed that he believed that he was really a human being and not a sheep at all. And even now, eight years later, when Nettle is a big grown up sheep, whenever Sophie goes into his field he comes

over for a scratch between his horns and a bit of a cuddle, just as he did when he was a little lamb.

The Soays that ran into the woods

Not long ago the lady with the largest flock of Soay in United States sold four young sheep called Retama, Tamarak, Madrone and Elm to a new owner who had a farm up in the mountains in the far west of America. The farm was surrounded by dense woods full of deep ravines and fast flowing rivers. Hardly had the four little Soays arrived at their new home than a terrible storm blew up. There were great rolling thunder claps and huge bolts of lightening struck the ground all around. The four little sheep became so frightened that they jumped clean over the fence and ran far up into the dense woods behind the farmhouse. But the four little

Soays didn't know that these woods were full of fierce cougars, coyotes and bears.

Seeing her new sheep run off into the woods, their new owner, Corrine, was very frightened. She was certain that they would never be able to find their way out of the dense woods and back to their new enclosure. Soon, Corrine thought, they will be caught by a bear or a cougar, or hunted down by a pack of coyotes, and eaten. How could she help them? The woods were so deep and dense that she could never hope to find them. And even if she did find them, how could she catch them and bring them back home? So, still not knowing what to do for the best, Corrine laid a trail of food from the sheep's field up as far into the woods as she dared. Perhaps the four little sheep might find it and follow it back home.

Two whole days passed but no little Soays returned. Time and again Corrine went to her back window and looked up towards the woods. But in vain. By now she feared the worst. All four must have been caught and eaten.   Then, early in the morning on the third day, Corrine heard bleating at the back of the house. Running outside she found little Retama and Madrone standing by the fence pawing at it. Overjoyed, she rushed to the gate and let the two little Soays back into the field.

Another eleven long days passed with still no sign of the

other two little Soays, Tamarak and Elm. Corrine was now certain that she would never see them again. But then, on the evening of the twelfth day, as she was looking out of her front window down to the road that led up to her house, Corrine let out a cry of astonishment. There, standing by the fence sniffing noses through the wire with Retama and Madrone, was Tamarak. She could barely believe her eyes. How ever had he managed to make his way all the way around through all the woods, over all the rivers and across all the ravines, and at the same time, for eleven whole days, dodge the cougars, coyotes and bears to find his way home to a strange new place all on his own? Quickly Corrine ran down to the bottom of her field and let Tamarak in.

Tamarak, one of the Soays that came back, and his friend QP.

For a long time after that Corrine would go repeatedly to each of her windows hoping to see her fourth little Soay. But to this day he has never returned. Corrine knows that he has probably long ago been eaten by one of the fierce beasts in the woods. Yet she still hopes that Elm may be out there somewhere, still dodging the cougars, coyotes and bears.

Could he be? What do you think?

If you want to learn more about Soay sheep you can see them at most rare breed farms or you can visit the Soay Sheep Society website at
www.soaysheepsociety.org.uk

Part of any profits from the sale of this book will go to The Soay Sheep Society to further its work of protecting and increasing knowledge about Soay in Britain and around the world.

Author's note

I am deeply grateful for the unstinting help given to me by Christine Williams and Kathie Miller, whose knowledge of Soay sheep far exceeds my own. As well as helping and encouraging me while I was writing this book they have provided many of the pictures. I also wish to thank my colleagues, Debbie Slater and Jeremy Beale at Boa Ms Limited, who not only prepared the typescript and brought the book to publication but have been unfailing in their help and advice. I also thank my wife Sophie for reading the many succeeding versions of my manuscript and offering many helpful suggestions. Most particularly, I thank Sam Green. Without his curiosity and enthusiasm I would never have started this book. Lastly, but by no means least, I have to mention our own small flock of Soay. For the last eight years they, Nettle among them, have been an ongoing joy and fascination.